# GEOLOGY AND FOSSILS OF THE HASTINGS AREA

**Ken Brooks**

**Acknowledgements**

I am very grateful to the following people for their assistance in preparing this book: Diana Williams (artwork), Tom Walton (computer graphics), Eric Robinson (Geologists' Association), Peter and Joyce Austen (Wealden plants), Angela Milner, Alison Longbottom, Alan Warren, Paul Davis, Phil Palmer, Paul Kendrick (Natural History Museum specialists), Ed Jarzembowski (Maidstone Museum curator), Julian Porter (Bexhill Museum curator), and John Cooper (Booth Museum curator).



Published by Ken Brooks
Printed by D C Graphics Ltd, Bexhill-on-Sea

ISBN 0 9540513 0 0

*Front cover:* Rock-a-Nore, Hastings
*Back cover:* Dinosaur footprints, Bexhill

# CONTENTS

# INTRODUCTION

It has taken many millions of years for the gradual processes of continental collision and erosion to produce the distinctive landscape that we see today in South-East England.

This book begins with a brief geological history of the Wealden district. This is followed by detailed descriptions of two field trips which will involve looking at fossils and sedimentary structures in the Hastings area. Then, by studying the evidence gathered at local sites, an attempt will be made to reconstruct the ancient environments and communities. To achieve this we must try to 'read' and interpret all the available clues, rather like a detective solving a mystery.

In 1991 the Nature Conservancy Council (now English Nature) designated the cliffs and beach from Rock-a-Nore to Pett as a Site of Special Scientific Interest. It concluded that some of the local fossils are the finest examples of their type in the world and that 'this is probably the best area for future finds of Lower Cretaceous reptiles outside of the Isle of Wight'.

# GEOLOGICAL TIME CHART

| Era | Period | Million years ago |
|---|---|---|
| | | 0 |
| QUATERNARY | Recent | |
| | | 0.01 |
| | Pleistocene | |
| | | 1.6 |
| TERTIARY | Pliocene | |
| | | 5 |
| | Miocene | |
| | | 23 |
| | Oligocene | |
| | | 35 |
| | Eocene | |
| | | 57 |
| | Palaeocene | |
| | | 65 |
| MESOZOIC | Cretaceous | |
| | | 145 |
| | Jurassic | |
| | | 208 |
| | Triassic | |
| | | 215 |
| PALAEOZOIC | Permian | |
| | | 290 |
| | Carboniferous | |
| | | 363 |
| | Devonian | |
| | | 409 |
| | Silurian | |
| | | 439 |
| | Ordovician | |
| | | 510 |
| | Cambrian | |
| | | 570 |
| | Precambrian | |

This chart shows the relative ages of sedimentary rocks in the Hastings area.

**Wadhurst Clay**
**Ashdown Formation**

They form the lower part of a group within the Wealden Series.

# THE LOWER CRETACEOUS WORLD

## ( 140 - 120 million years ago )

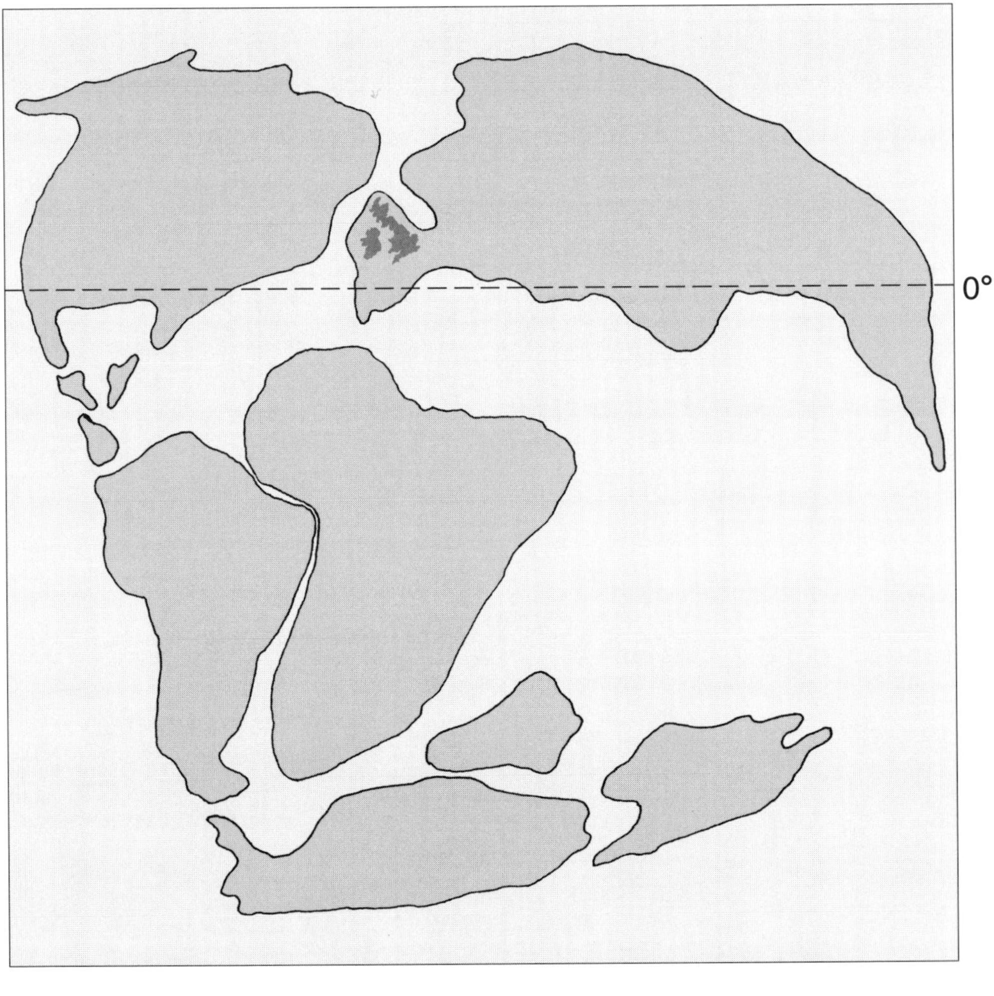


**Map with approximate position of Britain**

# GEOLOGICAL HISTORY

During the Lower Cretaceous period Britain was part of the European land mass. Fossil evidence indicates that southern England had a sub-tropical climate with seasonal rainfall, perhaps like the Mediterranean today! A rich variety of plants and animals lived on the marshy land and in fresh to brackish water environments, while rivers deposited huge quantities of sand, silt and mud in lakes and on flood-plains. The great weight of these sediments, combined with geological faulting, resulted in a gradual subsidence of the whole area.

Then, around 100 million years ago, marine life was established when a warm, shallow sea began to cover most of England and northern Europe. By the end of the Upper Cretaceous, 65 million years ago, the remains of countless millions of micro-organisms had formed thick beds of white sediment, which later hardened into Chalk.

Meanwhile, Africa was moving slowly northwards towards the European continent and, when the two land masses eventually came together, it caused intense folding of the Earth's crust. The Alpine mountain chain was formed and South-East England was pushed upwards into a huge dome-shaped structure (now known as the 'Wealden Anticline'). All of these events occurred over a vast time span that is almost impossible to comprehend.

Since then, erosion has removed the overlying beds in the central Weald to reveal the sandstones and clays of the Lower Cretaceous 'Hastings Group'. It is these ancient rocks that determine the shapes of the hills and valleys of South-East England today.

# FIELD TRIP LOCATION MAP

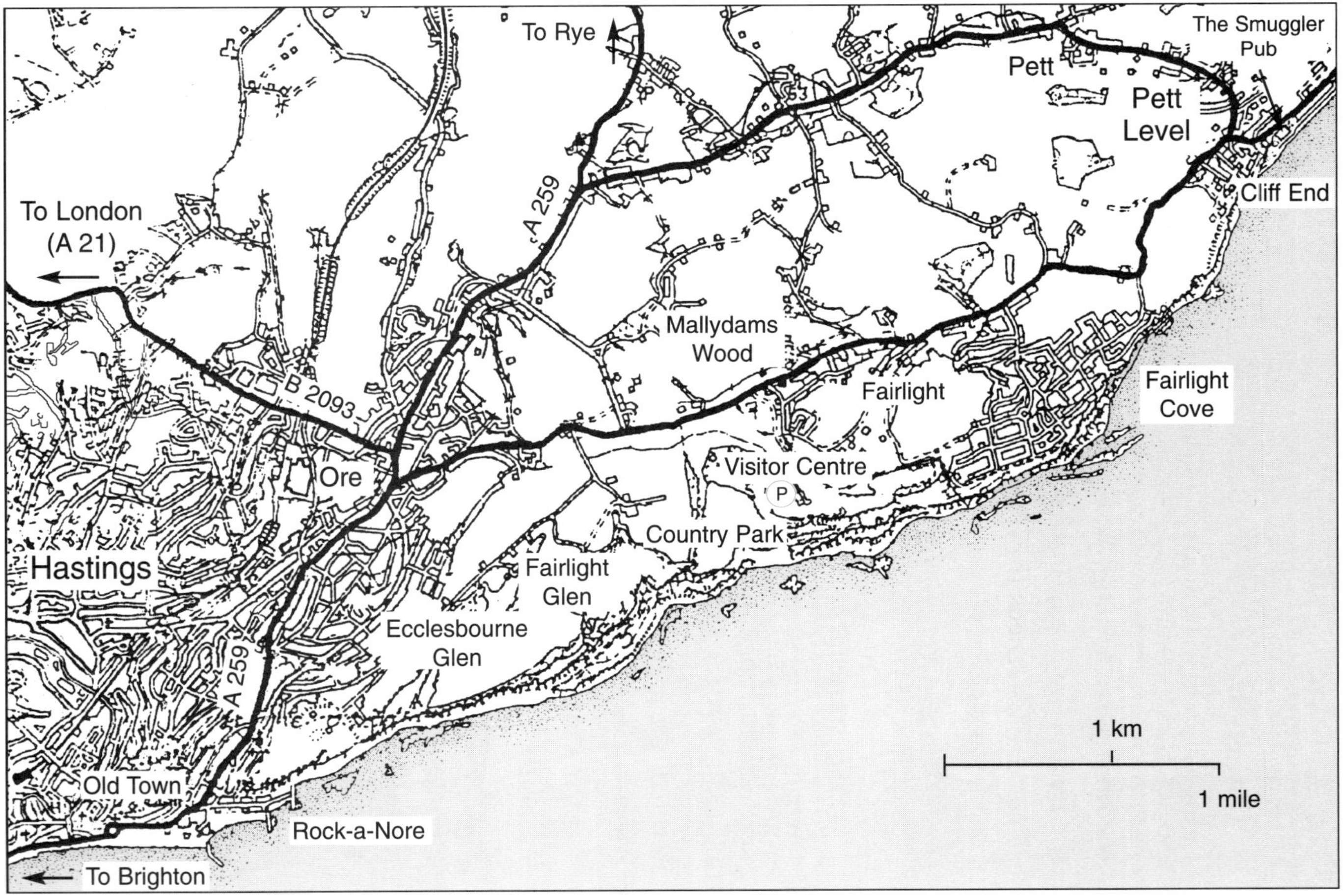

# FIELD TRIP NOTES

There are a few important safety rules that should be observed on a coastal field trip:

(1) Always check the tide times, and set off on a falling tide.
(2) Let someone know where you are going.
(3) Do not walk directly below the cliffs.
(4) Never attempt to climb the cliffs.
(5) Wear suitable boots or shoes, and a hard hat.

It is also useful to carry a First Aid kit, a mobile phone, waterproof clothing, a geological map of the area, a hand lens, camera, notebook and pencil. If your observations are to be of some value, the most important detail to record is the location of particular feature or fossil. Other information should include the date, a description of the rock and the identity of a specimen, if known.

Please do not remove fossils from the rocks, unless you intend to use selected examples for study. It is unfortunate that the same processes of erosion which reveal fossils will eventually destroy them but, until then, others may wish to see them.

# ROCK-A-NORE FIELD TRIP

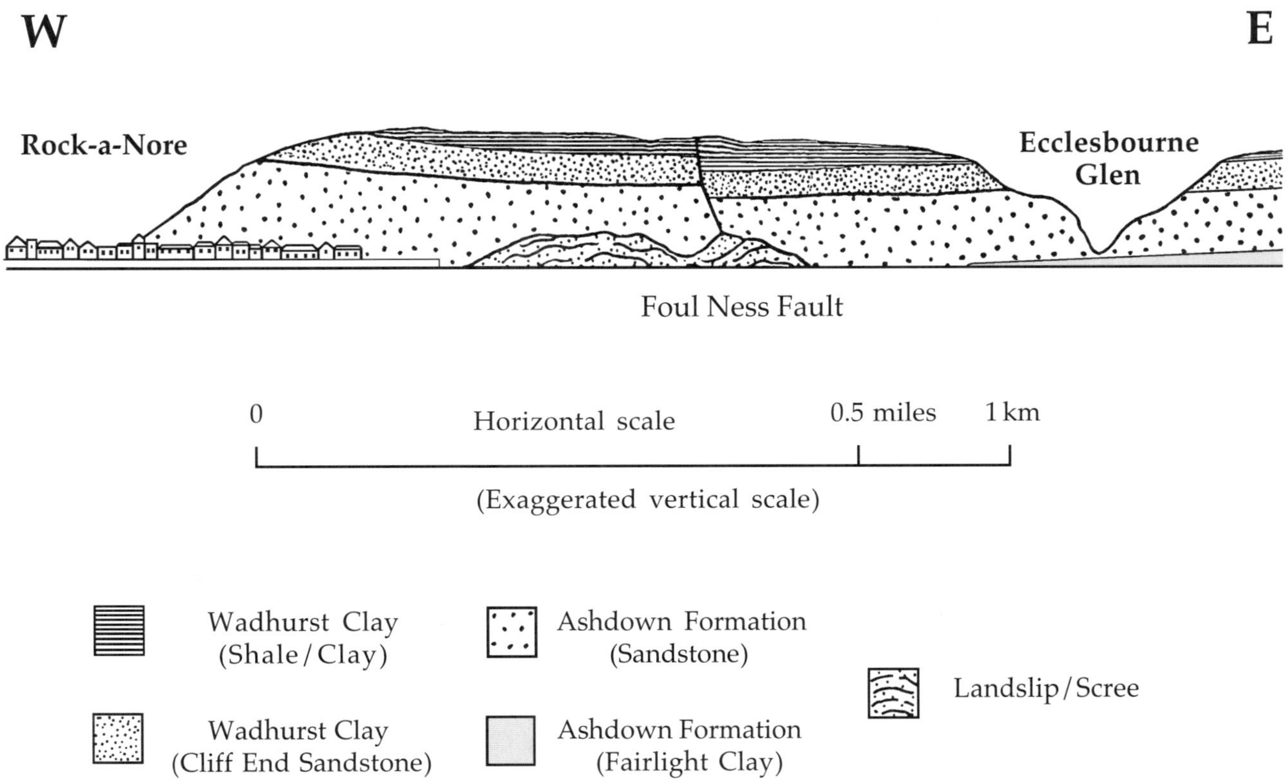

# ROCK-A-NORE FIELD TRIP

(O.S. map 199 B.G.S. sheet 320/321 Grid ref. TQ 831 095)

Follow the Hastings sea-front eastwards to the 'Old Town' and the famous net 'shops' in Rock-a-Nore Road. Below the high sandstone cliffs of the East Hill you will find the Fishermen's Museum, Underwater World (a Sea Life centre), a large car park and public toilet facilities.

This field trip begins at the last stone groyne and continues along the beach towards Ecclesbourne Glen, nearly 1 km ( half a mile) to the east. The massive sandstone cliffs of the Upper Ashdown Formation are overlain by shales and sandstones of the Wadhurst Clay. A distinct junction between one horizontal bed of rock and another often marks a period of erosion. This may have been followed by a change in the environmental conditions where a different grade or type of sediment was deposited.

In this area the lower part of the cliff is hidden under a scree slope of broken rocks, but there is one small exposure *in situ* at beach level. Here there are flattened branches of carbonised wood lying horizontally within a silty mudstone. These were probably washed into a river or lake, then later covered and compressed by sedimentary layers. The leaves of small ferns, such as *Weichselia*, may also be found at this location. These delicate plant remains are preserved as fossils only because they contain carbon - which is derived from the same carbon that was fixed by photosynthesis over 100 million years ago!

Fragments of the tree fern *Tempskya* occasionally fall from a layer near the top of the cliff. After the plant died it was covered by sediments and, over millions of years, its 'trunk' was gradually replaced by silica (quartz). In fact, with a hand lens, tiny crystals of quartz may be seen lining the spaces between a mass of brown-stained fibres. The outer part of a trunk section is sometimes covered by fusain, a friable layer which represents the carbonised remains of external tissues.

The fine-grained sandstones and siltstones occasionally contain specimens of a freshwater gastropod (snail) *Viviparus* (Plate 1). This is also the name of its present-day descendant which lives in ponds and streams. Isolated specimens of the fossil *Viviparus* may sometimes appear among shells of a small bivalve. This bivalve, *Neomiodon*, is the most common of all fossils in the Hastings area. The bedding planes of sandstones are often covered with *Neomiodon* shells (Plate 2), and layers of them can be seen in cross-section on the sides of blocks. There are also larger bivalves which may be found locally, such as *Filosina* and *Unio*, but these are less common.

Small tube-like and oval structures are often found in rocks near the shells of *Neomiodon*, and these have been identified as its burrows and feeding pellets. Where the sediment has been disturbed or burrowed by any animals, it is known as 'bioturbation'.

Trace fossils, such as footprints and burrows, sometimes provide valuable information about an animal's presence and behaviour, particularly where no other evidence has been found. However, the nature of the species itself can be difficult to deduce, as different organisms may have had a similar mode of life. Some siltstones have preserved the remains of conchostracans, small crustaceans which are often described as 'clam shrimps' (Plate 3). These were enclosed in shell-like carapaces, usually measuring about 7 mm. (0.25 in) in length and, because of this, their fossils are sometimes mistaken for bivalve shells.

Even smaller are the oval-shaped ostracods *Cypridea* (which are also crustaceans) but, as they measure only about 1 mm (0.04 in) in

Plate 1. Gastropods: *Viviparus*

Plate 2. Bivalves: *Neomiodon*

Plate 3. Conchostracans

Fish teeth: *Lepidotes*
Plate 4.

Crocodile tooth: *Goniopholis*
Plate 5.

length, a hand lens is required to examine them closely. Although their tiny seed-like carapaces can be found in large numbers on the bedding planes of thin shales, it is easy to overlook them on a field trip. Some species of *Cypridea* have ornamented carapaces with raised 'pimples', while others have smooth surfaces.

The siltstones also contain evidence of the insects that lived around the streams and lakes. For example, the larval tubes of caddis flies (Trichoptera) may be found. The larvae of these insects used locally available materials to build the tubes which would serve as temporary homes while they lived in the water.

Fish and reptile remains are usually found as isolated fragments, probablt because they were broken up and dispersed by streams or rivers. These fossils include the glossy, brown scales and small rounded teeth of the fish *Lepidotes* (Plate 4), and the pointed teeth and fin spines of the shark *Hybodus*.

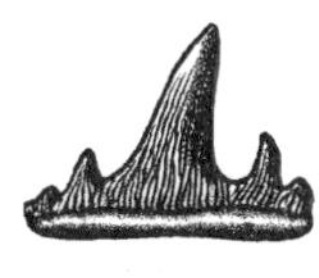

*Hybodus* tooth

1 cm

Individual scales from *Lepidotes* are common (Plate 6), but to find an undamaged specimen of this fish is rare. Part of a body with well-preserved scale impressions is still visible in a hard sandstone block near Rock-a-Nore, and a few years ago a complete *Lepidotes* was discovered at Covehurst Bay (near Fairlight).

There are also the scutes (horn-like plates) from turtle shells and the bones and teeth of crocodiles, such as *Goniopholis* (Plate 5) and *Bernissartia*. Turtle shell scutes can be smooth or covered with raised 'pimples', while the plates from crocodiles' skin have circular dents in them. These are sometimes exposed in iron-stained 'pebble beds' or on the bedding planes of a friable sandstone. This rock's crumbly surface is the result of disruption by salt crystallisation from sea water, and freshly weathered layers of a particluar bed frequently contain fossil specimens.

Some of the rarer fossils from this area include the phalanges (finger bones) of the pterosaur *Ornithochierus*, one of the largest of all the flying reptiles. Such specimens may be identified as pterosaur remains, as the once-hollow bones which allowed them to fly are usually crushed flat.

Perhaps the most impressive finds are the teeth and bones of dinosaurs. These can occasionally be discovered on the beach or at the foot of a scree slope

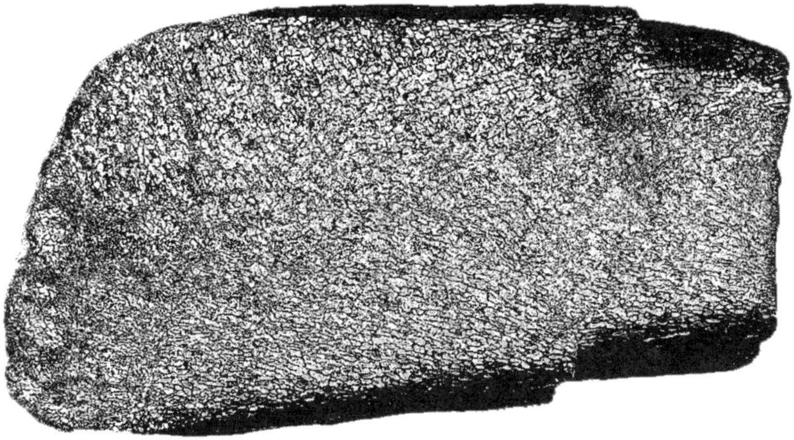

Section of dinosaur bone 1 cm

More common are the large tridactyl (three-toed) footprints of dinosaurs, such as *Iguanodon*, which may be seen on the bedding planes of certain siltstones (Plate 7). Some of the prints can measure up to 60 cm (2 ft) from heel to toe. These dinosaurs might have been walking by a river or lakeside and left their footprints in the mud. This quickly dried out in the hot sun, and water then deposited silt or sand in the hardened impressions, later forming infill casts. Gradually, over millions of years, these sediments turned into rock. Today either the footprint impressions or the casts may be revealed when the weaker rock layers are removed by erosion.

It is a fascinating thought that, while fossilised shells, bones and teeth are natural reproductions, a dinosaur footprint is the actual impression left in the mud millions of years ago by a living animal!

Slabs of a bluish-grey rock with strange bulbous shapes may be seen on the beach in this area. These are fallen blocks of Tilgate Stone which come from various strata within the Wadhurst Clay. In the past this rock was known by local quarrymen as 'Bluestone' or 'Hastings granite'. It is a hard, fine-grained sandstone which is strongly cemented by calcium carbonate (calcite) into tabular and nodular masses (Plate 8). The Tilgate Stone's unusual dome-shaped structures may have been formed by the attraction of calcium carbonate from groundwater around a tiny fragment of shell or a crystal of calcite. The concentric hardening of the rock gradually produced the rounded structures (concretions) that we see today, now that the poorly cemented outer layers have been removed by erosion.

About half a kilometer (quarter of a mile) to the east of Rock-a-Nore the horizontal strata of the cliff are displaced by the Foul Ness Normal Fault. However, this is obscured by a section of cliff which has slipped down to form a small plateau. Below this is a scree slope of broken rocks and shale where segments of the horsetail *Equisetites* may be found in mud slides on the beach.

These specimens usually measure between 3 and 8 cm (1 to 3 in) in length. They can be recognised by their 'striped' appearance and tiny holes where leaves once grew around the stems.

*Equisetites lyelli*

1 cm

Other interesting discoveries which have been made in this area include fossil bones and an *Iguanodon* tooth.

From here to Ecclesbourne Glen most of the fallen rock and scree material lies directly below vertical cliffs and, for safety reasons, close examination is not recommended. However, there are still features of interest in scattered rocks along the foreshore, and even the shingle itself is worth closer study.

It is important to realise that the flint pebbles covering the beach do not originate from this area. The sea has washed them along the coast from Beachy Head, where the high Chalk cliffs contain many nodules and thin beds of flint. Between 100 and 65 million years ago, Britain was covered by a shallow sea where sponges and other marine animals extracted silica from the water to strengthen their fragile structures. Later, over a period lasting many millions of years, their skeletons dissolved and the silica eventually recrystallised as concretions of flint. It should be pointed out that not all geologists agree with this explanation and that the chemical processes involved in the formation of flint may be far more complex. For example, the silica could have been precipitated directly from sea water.

Heart-shaped echinoids (sea urchins) are among the fossils that may be found in the shingle, but more common are the hollow flints that once enclosed delicate sponges. These often contain very small and beautiful crystals of quartz (chalcedony) which are best seen with a hand lens.

Ecclesbourne Glen is an ancient river valley. It was carved out by vast quantities of melt-water which flooded across southern England at the end of the last ice age, around 10,000 years ago. Since then, the sea has gradually cut back into the cliff to form a hanging valley with a small waterfall about 6 m (20 ft) above the beach.

This location marks a suitable turning point for walking back to Rock-a-Nore.

# GEOLOGICAL FAULTS

## NORMAL FAULT

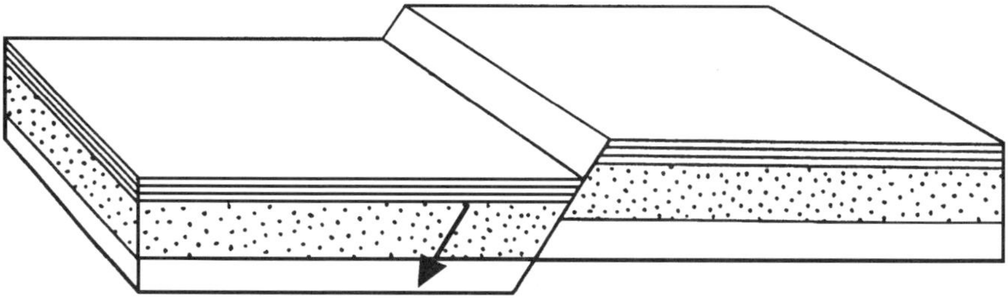

Tension ←  → Tension

**Extension of surface rocks**

## REVERSE FAULT

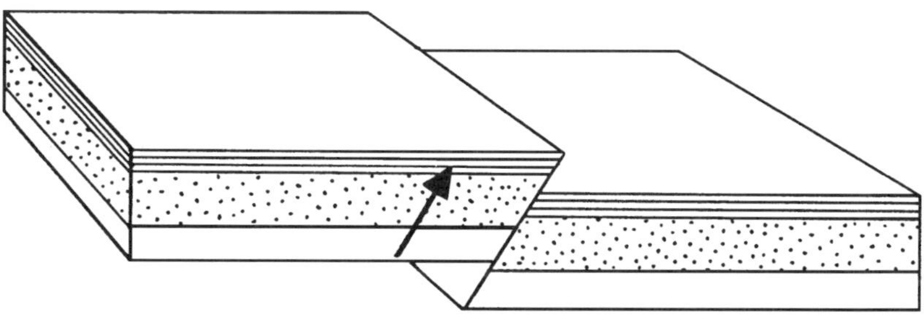

Compression →  ← Compression

**Compaction of surface rocks**

# PETT LEVEL FIELD TRIP

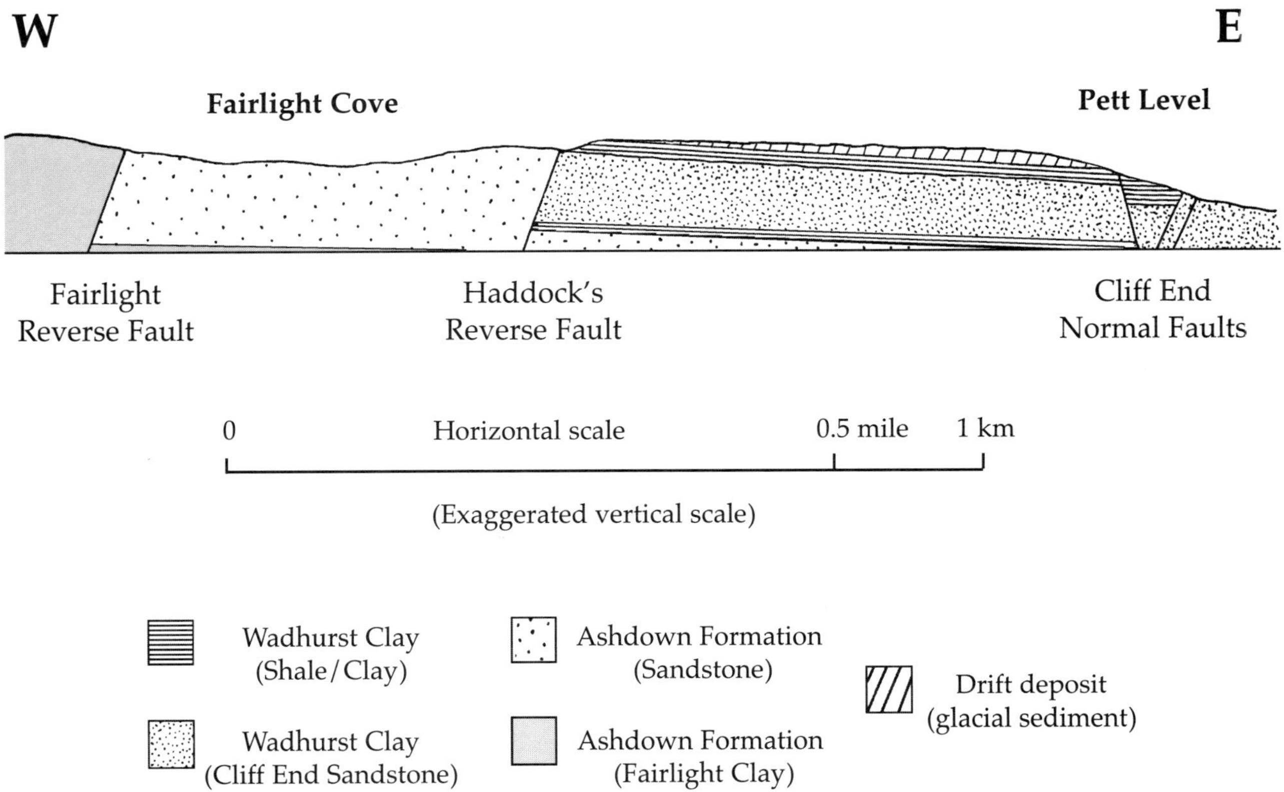

# PETT LEVEL FIELD TRIP

(O.S. map 199   B.G.S. sheet 320/321   Grid ref. TQ 890 133)

This field trip begins at the Smuggler Pub, where cars may be parked on the opposite side of the main road. There are public toilets about 50 m (55 yds) along the road to the west. Walk towards the beach, turn right and follow a concrete promenade which leads to a small bay enclosed by sandstone cliffs. These are formed from Cliff End Sandstone, a unit of about 10 m (33 ft) thick within the Wadhurst Clay. At this location the cliff is displaced by 'normal' faults which were caused by tension (rock stretching) during the Wealden uplift. Geological faults are often obscured by eroded material, but here there is a text-book example of a normal fault which clearly shows the slip plane and downthrown strata (Plate 9). The sandstone near the top of the cliff is stained grey or black by carbonised plant material, possibly deposited as fusain (charcoal) from forest fires. The rock also contains traces of vertical roots extending downwards. This quantity of plant material suggests that the sand may have settled on a delta or flood-plain, with rich vegetation growing nearby.

In the upper part of the cliff, to the left of the main fault, there is a small cave which was probably eroded by melt-water at the end of the last ice age. Recent freeze-thaw action has produced vertical fissures where huge columns of sandstone are split from the cliff face. In fact, the cliffs along this part of the coast are particulary unstable, and rock falls can occur without warning.

When the tide is out, the remains of a 'submerged forest' may be seen in the mud and sand of the beach. The wood is soft and black, but it is possible to identify birch branches, hazel nuts and tree roots, still in their original growth positions. Carbon dating has shown that these trees were growing here around 5000 years ago, when the sea level was lower and a wooded area extended from Pett Level to Bexhill.

Continue westwards along the shingle towards Fairlight Cove. Near the top of the cliffs the Wadhurst Clay shales contain thin lenses of a very coarse sandstone, known as the Cliff End Bone Bed. Broken fragments of this fossiliferous rock may be found on the beach at a safe distance from the foot of the cliff. The bone bed is composed of large grains of quartz with nodules of clay-ironstone (siderite) and many small teeth from the fish *Lepidotes* and the shark *Hybodus*. There are also bivalve shells, carbonised plant fragments, fish scales, crocodile teeth, worn pieces of bone and very rare rodent-like teeth of the earliest mammals. Although it has been described as a flash flood conglomerate or a river overflow deposit, the Cliff End Bone Bed is probably a concentration of material which settled as the finer particles were washed further downstream.

Below the Cliff End Sandstone is the basal Wadhurst Clay Ironstone, a rock which outcrops in other areas throughout the south-east. In mediaeval times it was one of the most important sources of iron-ore for the Wealden iron industry.

Some slabs of fine-grained sandstone contain sedimentary structures preserved on bedding planes. These include sand ripples which were shaped by flowing water on a river bed or by oscillating wave currents at a lakeside.

Positive casts of narrow water channels, known as gutters, are also quite common (Plate 10), although they are sometimes misidentified as fossilised tree roots or crustacean burrows. It may be noted that the 'snake-like' structures always trend in one direction, and that small shells are sometimes found where they sank to the base of a channel. These observations provide clues to the true origin of the casts. They were probably formed when flowing water carved thin grooves on a flood-plain or river estuary, and these filled up with sediments which eventually hardened into stone .

There are certain siltstones which have a 'churned-up' appearance where a number of *Iguanodon* have left their footprints and toe impressions. Heavy animals, such as these dinosaurs, must have caused considerable disturbance in muddy sediments ('dinobation'). Other features in this particular bed of rock include curved grooves which have been interpreted as dinosaur tail drag marks (?)

After winter storms it is possible to find dinosaur bones that have been washed from shales and siltstones under the shingle or from fallen cliff material. The most recent discoveries in this area include a turtle's carapace, dinosaur vertebrae and a rib measuring approximately 80 cm (2 ft 6 in) in length. Fossil bones are often stained brown by groundwater containing dissolved iron, which may be derived from the sandstones.

The beach near Haddock's Reverse Fault is the location of several small, three-toed footprints on a bedding plane. The impressions, which average 15 cm (8 in) from heel to toe, are much more bird-like than *Iguanodon* footprints. Nearby there is a slab of fine-grained sandstone with three positive casts of similar footprints. These might have been made by an unidentified species of theropod (carnivorous) dinosaur!

With the exception of horsetails, fossilised plants are rarely found in their life positions, possibly because rivers and floods frequently eroded and removed material during the Lower Cretaceous. However, despite these conditions, the first example of an *in situ* Wealden quillwort, *Isoetites*, was recently found here in a sandstone ledge. Unfortunately, these quillworts are preserved in a weakly cemented rock, and their iron stained outlines are soon destroyed when exposed to present-day erosion by the sea.

At this location the cliff reaches a height of 20 m (66 ft) and there are several caves at beach level which were carved out by the sea.

Further along the beach there is a small promontory with the exposed plane of the Haddock's Reverse Fault on the western side. This fault was the result of massive lateral compression, when the older Ashdown Sandstone was forced upwards against the younger Cliff End Sandstone. At some later time part of the cliff slipped back down along the fault plane. Within the fault plane, and at other places along the base of the cliff, there are some interesting sedimentary structures. Among these there are several wedge-shaped 'point bar' deposits. These formed when sediments built up on the inside of a meandering river bend where the water flow was slower. What we now see is a cross-section of a river bend with each thin layer of sandstone representing a period of deposition.

There are also cross-sections of curved beds filled with thin layers of sandstone (Plate 11). These are channels which were were probably carved out by fast-flowing streams and, as the current speed decreased, sands and silts were deposited within them. On the beach opposite Haddock's Reverse Fault is a large outcrop of iron-stained sandstone containing carbonised rhizomes and stems of the horsetail *Equisetites,* still in their original growth positions. When the plants died, fine-grained sediment filled their hollow stems, thereby ensuring their preservation. The top surface of the sandstone contains dark spots which are cross-sections of stems, while the sides show the outlines of the shoots as they grew upwards through the sediments (plate 12).

From here we can see the bay of Fairlight Cove with its abandoned houses perched on the cliff edge and a barrier (or berm) of huge rocks extending along the beach.

The rapid erosion of the cliffs in Fairlight Cove is the result of geological faulting during the Wealden uplift. Where the soft clays and silts of the Lower Ashdown Formation are exposed at beach level they are easily worn away by the sea and, without underlying support, the sandstone cliffs above collapse on to the beach. When fallen rocks and shingle are washed away, this leaves the base of the cliff quite unprotected, and so the process continues.

During the past 100 years the Fairlight Cove bay has been eroded at an alarming rate, until the homes of local residents were literally falling over the edge of the cliff. It has been calculated that the average rate of erosion along this stretch of coast is about 1 m (3 ft) per year.

In 1990, after a determined campaign by the Fairlight Coastal Protection Association, the local authorities agreed to build a sea barrier below the cliffs. Huge boulders of larvikite were transported in barges from Norway and placed on the beach between the Haddock's and Fairlight Reverse Faults. (Larvikite is a granite-like, igneous rock which is identified by its beautiful, iridescent blue/grey feldspar crystals). The scheme allows cliff falls to continue until the rock material forms a stable slope which will be protected from the sea by the barrier. From recent observations it appears that the plan is working because, as the rate of erosion has decreased, vegetation is now growing on a well-established scree slope.

On the seaward side of the rock barrier there are beds of silty clay which contain a number of *Iguanodon* footprints, although they are often covered by sand and shingle after stormy weather. It has been suggested that, as some of the prints appear to form track sequences, they might represent a feeding ground route for *Iguanodon* herds (?)

Another feature of interest is the small river channel which can be seen on the foreshore at this location. Its gently curving bed clearly cuts through dipping layers of siltstone where it apparently flowed southwards over 100 million years ago.

About 60 m (200 ft) past the western end of the sea barrier is the Fairlight Reverse Fault, which is now covered by a landslip. However, the site is marked by an abrupt change from vertical sandstone cliffs of the Upper Ashdown to the sloping clays of the Lower Ashdown, known locally as the Fairlight Clays. These are easily identified by the yellow, purple and red-staining which was produced by the oxidation of dissolved iron minerals in groundwater.

The Fairlight Clays were once famous for the variety of ferns preserved in them, but it is now very difficult to find good specimens. Today the most common plant fossils include stem fragments of *Equisetites* and small fronds of *Onychiopsis* and *Weichselia*.

Beyond this point it might be worth looking for fossils in fallen blocks of clay before returning eastwards to Cliff End.

*Onychiopsis* 2 cm

# ENVIRONMENTAL RECONSTRUCTION

The following reconstruction is based on the study of rock structures and fossils from the Hastings area and other locations of the same geological age.

## The Physical Environment

Particular types of sedimentary rocks can tell us much about the environment at the time of their deposition. This is because different forms of sediment were laid down according to various conditions in the water. These depended upon the type of sediment carried by a river, the depth of the water and the sorting action of changing currents. Many local sandstones were formed when rivers deposited thin beds of sand with the grain size of each layer determined by the energy (flow speed) of the water. Gradually, the layers built up to produce sedimentary structures known as 'graded bedding'. For example, coarse-grained and heavier sand would be deposited in turbulent water, while the finer material was carried further down-stream. The silt and clay particles eventually settled from suspension to form sediments in the calmer water of an estuary or the deeper parts of a lake. Overflowing rivers deposited silt, clay and mud on their flood-plains, while sediments also built up near the mouth of a river, perhaps forming mudflats or a prograding delta into the lake.

Even the particles which form local sedimentary rocks can tell us something of their origins. For example, the mica in clay and the quartz in sandstone are two of the most common minerals in granite. Very ancient granitic rocks must have been broken down by erosion, transported by water, deposited as sediments and then consolidated into new rocks. This recycling process may have occurred several times over many millions of years before the clays and sandstones of South-East England were formed.

Many of the sandstones are formed from well-rounded grains which were shaped by collisions with other fragments in moving water. Their degree of roundness is evidence of transport over a considerable length of time and distance, or they could be on their second or third time around as recycled rocks. Poorly sorted beds usually contain large and small angular fragments which were quickly deposited by floods or fast-flowing rivers. Rocks with rounded particles of varying sizes, such as the Pett Bone Bed, were sometimes formed as a result of winnowing (sorting).

As meandering rivers flowed across the landscape they left their distinctive patterns in the sedimentary record (Plate 11). On the outer sides of river bends the curves were eroded and widened by fast moving water. At the same time sand and silt settled from the slower water on the insides of bends to form point bar deposits. It was this process that increased the snake-like meanders of a river and enabled it to spread sediments over a wide flood-plain.

Occasionally, perhaps after heavy rainfall had increased the flow velocity of a river, part of a sedimentary sequence was washed away, leaving layers which were cut off quite abruptly. A return to calmer conditions and renewed deposition is marked where the base of this later sediment cuts across the earlier laminations.

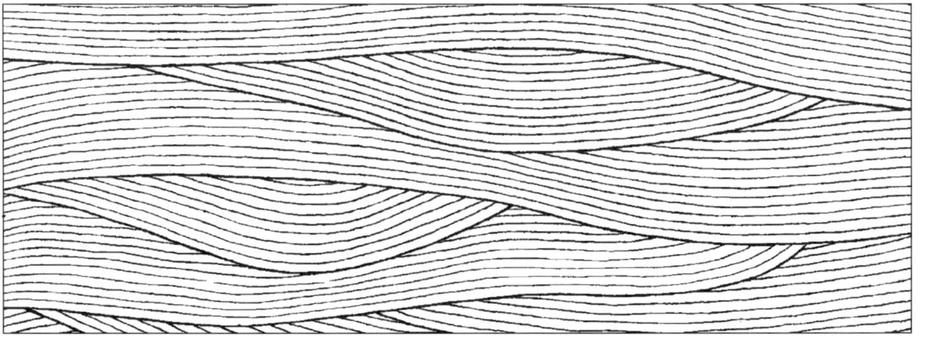

Truncated laminations 5cm

Structures such as these illustrate how difficult it is to calculate a period of time from the thickness of a sedimentary sequence. This is because we do not know how many layers were removed by erosion between each deposition phase. However, truncated laminations are very useful in determining the original way up of a fallen block of sandstone, as the more recent layers always cut across the older beds.

The remains of fish, reptiles and plants were swept downstream and deposited in calmer water where they were covered by sediments. Seasonal floods left pools of shallow water which would then become stagnant in the dry season. These sometimes provided ideal conditions for the preservation of plants and animals as, without oxygen in the water, there would be no aerobic bacteria or scavengers to destroy organic material.

The fine-grained nature of the Fairlight Clays suggests that these sediments were deposited in slow-moving or deep water. Later, the sandstones of the Upper Ashdown Formation appear to have been laid down in braided streams and meandering rivers on flood-plains. These massive beds of sandstone were probably formed as a result of uplift in the London area, where renewed erosion by rivers (rejuvenation) brought coarser material to the flood-plains and lake deltas of the south-east. As rivers flowed into lakes they deposited sediments to form deltas, while the thickness of the beds suggests a steady supply of land-derived material into a subsiding area over a very long period. During the rainy season much of the region would have been flooded, allowing layers of silt and sand to settle in horizontal beds. As the floodwater drained away it left many small water channels (gutters) in the soft, muddy silts (Plate 10). Sand ripples and plant roots *in situ* clearly indicate shallow water conditions, but perhaps the most dramatic evidence is provided by dinosaur footprints.

# KEY TO RECONSTRUCTION

## ( Centre pages )

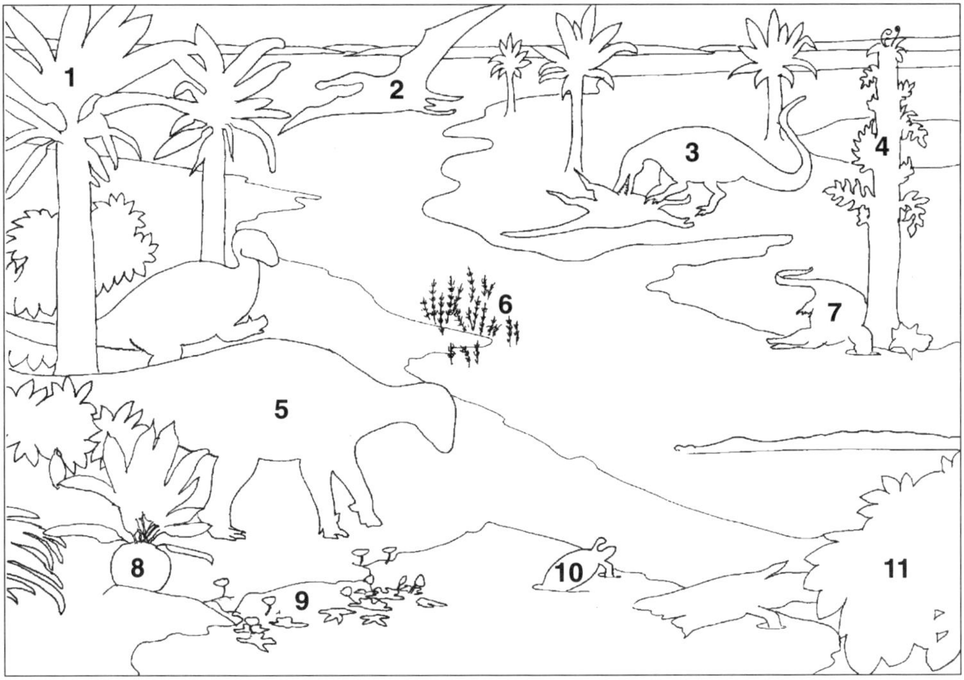


1. Bennettitalean: *Monanthesia* sp.
2. Pterosaur: *Ornithocheirus* sp.
3. Dinosaur: *Baryonyx walkeri*
4. Tree fern: *Tempskya schimperi*
5. Dinosaur: *Iguanodon* sp.
6. Horsetails: *Equisitites lyelli*
7. Crocodile: *Goniopholis crassidens*
8. Bennettitalean: *Cycadeoidea* sp.
9. Flowering plant: *Bevhalstia pebja*
10. Turtle: *Plesiochelys* sp.
11. Fern: *Weichselia* sp.

Plate 6. Fish Scale: *Lepidotes*

Plate 7. Dinosaur footprint: *Iguanodon*

# IN LOWER CRETACEOUS TIMES

Plate 8. Sandstone concretions

Plate 9. Normal Fault, Pett Level

In the absence of any fossils of marine organisms it appears that the area was completely cut off from the sea while the sediments of the Ashdown Formation were being laid down. However, the fossils of the overlying Wadhurst Clay indicate that the freshwater lake environment was changing to one of brackish water lagoons. This was possibly the beginning of a gradual rise in sea level which eventually flooded all of southern Britain during the Upper Cretaceous.

**Plants**

The low-lying flood-plains, deltas and lakesides provided ideal environments for a wide range of vegetation.

Among the most abundant plants at this time were those which reproduced by means of spores (pteridophytes), such as the ferns, tree ferns, horsetails and clubmosses.

Complete fern fronds are rarely found, as most plants soon decomposed in the humid conditions or were washed away and reduced to fragments in turbulent waters. The few plants which have been preserved intact were probably covered quite rapidly by sediment in quiet water.

Among the most common ferns of this area was *Onychiopsis*, with delicate fronds that grew to about 40 cm (16 in) in length. It was a low-growing plant which thrived in shady and moist conditions. Like other true ferns, the spores of *Onychiopsis* developed on the undersides of particular fronds. As its modern descendant, *Onychium*, has changed very little, perhaps we can deduce that *Onychiopsis* lived in a similar habitat.

*Weichselia* was a rather different type of fern as it grew large whorls of umbrella-like fronds. The carbonised remains of this plant often appear to have been burnt to charcoal (fusain) in undergrowth fires. It is possible that these fires were started by lightning, or even by internal combustion within rotting vegetation. *Weichselia* may have adapted to its environment by producing soral clusters (pods) which required heat to release the enclosed spores. A similar process occurs today with 'fire-climax' species, such as the conifers.

The horsetail, *Equisetites*, grew in abundance on the marshy ground and mudplains by lakes and rivers, and in appearance looked very similar to the present-day horsetail *Equisetum*. Their system of runners (rhizomes), which grew underground from the parent plant, enabled communities to spread rapidly. These sent up vertical shoots or stems with circles of spiny leaves growing on them. *Equisetites* reproduced by means of spores which were contained in cone-like structures at the tips of their shoots. During dry periods the horsetail stems would shrivel, and then regrow from the rhizomes when conditions improved. The length of vertical stems preserved *in situ* is possibly the result of rapid burial by flood sediments. (Plate 12).

Like others in the pteridophyte group, the quillwort *Isoetites* also reproduced by means of spores. Fossils of this species have been found in their life positions, complete with short, upright stems, leaf bases, corms and radial rootlets.

Although *Tempskya* is classified as a tree fern, it is not in fact a tree. Its false 'trunk' was simply a mass of tangled frond stems and rootlets. This enabled the plant to grow above its competitors, perhaps reaching up to 5 m (16 ft) in height. Narrow, tube-like holes in the fibrous trunk conducted moisture and nutrients to its delicate leaves. These grew from the sides of the plant, unlike modern tree ferns where leaves grow from a central point at the top (an apical 'crown'). *Tempskya* may have used molecules of silica from the soil to reinforce its stems during growth. This might later have contributed to its preservation as a fossil by attracting other molecules of dissolved silica from groundwater.

A stem section from another species of tree fern, *Protopteris,* was found in the Hastings area a few years ago. The sandstone cast, measuring 20 cm (8 in) in length and 8 cm (3 in) wide, was covered with a pattern of diamond-shaped scars which were left when leaves were shed. The fossil evidence indicates that these tree ferns lived in marshy habitats which were similar to those of the Everglade swamps of Florida today. Another ancient group of plants is known as the gymnosperms ('naked seeds') because reproduction was by means of seeds within cones (i.e. not contained in an ovary). These included the cycads and conifers.

A small fossilised cycad cone was recently discovered in sandstone on a local beach. Cycads looked like small palm trees, but they were not related in any way. They each had a short, unbranched trunk, with a crown of large palm-like leaves which were tough and leathery. All cycads had separate male and female plants, with their reproductive structures held in large cones.

A few cycad genera have survived to the present day in tropical and sub-tropical regions. They are almost identical to those of around 200 million years ago and are often described as 'living fossils'.

Conifers are represented locally by carbonised branches of various sizes, and these are common in the silts and sandstones between Rock-a-Nore and Pett. During forest fires in the hot and dry seasons, the resin produced by these conifers would have provided a highly flammable fuel. Small samples of fossilised resin have been found in the area, but so far no insect remains have been detected inside them!

The bennettitaleans belonged to an extinct group of gymnosperms which resembled cycads and possessed reproductive structures that looked more like flowers than cones. A few years ago a leaf from one of these rare plants, *Pterophyllum,* was found in the sandstone at Rock-a-Nore (Plate 13). Many palaeobotanists believe that bennettitaleans were close relatives of the flowering plants, although there is no direct evidence that they were their ancestors.

For millions of years the gymnosperms had survived with their tough stems and leaves, but it was during the Lower Cretaceous that the first true flowering plants, or angiosperms ('covered seeds'), began to spread and diversify. In order to reproduce more successfully than the gymnosperms, the angiosperms colonised the more open habitats alongside rivers or lakes, and developed more efficient methods of seed distribution. The soft leaves and fruits of these plants may have attracted herbivorous dinosaurs. Although there is no evidence to support this, the new flowering plants might have achieved a wider dispersal of their species - with seeds able to survive in the stomach of a dinosaur!

The first angiosperms had rather insignificant flowers with small petals. A recently discovered specimen of one of these, *Bevhalstia,* was found in the Weald Clay of Surrey. This low-lying plant grew on marshy ground or in shallow, fresh water and would have been easily accessible to browsing herbivores.

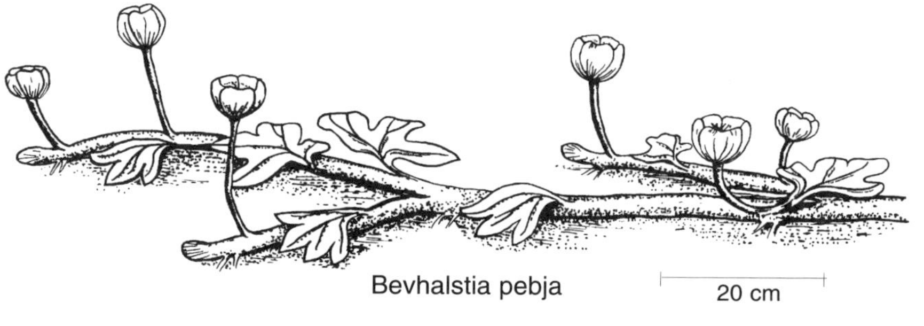


Bevhalstia pebja

From the evidence provided by fossil plants in the south-east it appears that the summers were hot and dry, followed by wet and humid conditions in the winters.

**Molluscs (shellfish)**

Molluscs extract calcium and carbonate from water to build (or secrete) protective shells around their soft bodies. These shells have concentric growth lines, showing where they were enlarged by the animal as it grew.

The most abundant mollusc in the Hastings area was the small bivalve *Neomiodon,* with an average measurement of 1 cm (0.4 in) in length (Plate 1). Its body was protected within two hinged valves/shells, with adductor muscles which contracted to close the shells. *Neomiodon* lived in the fine-grained, silty sediments of river estuaries or deltas and it may also have been adapted for life in brackish water.

Trace fossils, such as burrows and mud pellets, show that Neomiodon was a sediment feeder. It would burrow into the sediment to extract nutrients from the silt or mud, which was then deposited as oval-shaped pellets on the surface. Burrowing was also a very good technique for avoiding hungry predators. The length of their burrows, which are preserved *in situ,* indicates that *Neomiodon* could travel some distance in this way. Although this bivalve lived in fine-grained silts, its shells are often found in sandstones. It is possible that these were removed from their natural habitat by an increase in the current velocity of the water.

The great numbers of *Neomiodon* shells, which are seen as distinct layers on sandstone bedding planes, do not necessarily reflect large populations. It may be observed that most of the shells appear to be of about the same size. This suggests that, perhaps during violent storms, a large number of bivalves was washed away. Then, by a winnowing process, the heavier shells settled from the swirling water, while smaller specimens were carried further and deposited as the current energy decreased.

Another explanation would have involved a catastrophic event. For example, there might have been a sudden change in the chemistry of the water, or a community of bivalves could have been isolated in a pool during the dry season. In these cases we would expect to see paired bivalve shells of varying sizes. However, as most of those found are single valves of uniform size, it would appear that water transport was involved in their preservation.

Other bivalves, such as *Filosina* and *Unio*, would extract nutrients from the water by opening their shells so that fine filaments in the gills could trap floating particles of food.

The gastropod *Viviparus* was a different kind of mollusc (Plate 2). It had a spiral-shaped shell, and would have lived in a stream bed or in the shallow part of a lake. Like modern snails, *Viviparus* obtained food by scraping algae from rocks or other surfaces with its radula (a rasp-like tongue). As the soft-bodied animal grew bigger it enlarged the open part of its shell with calcium and carbonate from the water.

**Insects**

The larvae of caddis flies (Trichoptera) made their tube-shaped portable homes from minute fragments of plants, fish scales and conchostracan shells, which they bonded together with silk. The hardened outer cover would have provided an effective barrier against small predators, while inside it the larvae could pupate safely. The preservation of these larval tubes in fine-grained sediments indicates that the caddis fly larvae lived in the slow-moving waters of freshwater streams.

Various insect remains found at Fairlight, Cooden Beach and other locations in the south-east include the wing cases of cockroaches and fragments of grasshoppers, stoneflies, cicadas and dragonflies. In 1986 an almost complete specimen of a hawker dragonfly was found in a

Weald Clay pit in Surrey. It had a wingspan of 8.5 cm (3.5 in), and was preserved in very fine detail. True, or two-winged, flies (Diptera) are among the diverse orders of insects which may be found in the Wealden rocks.

### Crustaceans

Conchostracans were small, shrimp-like crustaceans that lived in shallow water or temporary pools. Each conchostracan was enclosed in a protective carapace formed from two hinged, shell-like 'valves' (Plate 3). Although its gill-bearing legs and antennae were exposed while swimming or feeding, these could be withdrawn into the carapace when danger threatened. Modern conchostracans are almost identical with their fossilised ancestors.

Ostracods, such as *Cypridea,* were similar to the conchostracans, but they were much smaller and more abundant. They had four pairs of legs which were used for swimming, crawling or burrowing, and the males would clasp females with their front legs during mating. As with other crustaceans, ostracods moulted many times before reaching adult size, leaving their empty carapaces in the muddy sediment. Their food, which included algae and diatoms, was obtained by both filter and deposit feeding, while they in turn would have been a very important source of nutrients for small animals near the bottom of the food chain.

The closely packed remains of these ostracods are usually found in thin shales, perhaps indicating that they lived in isolated pools which may have dried up rapidly in the hot season. We should also consider that the carapaces were probably the moults of a small number of ostracods over a period of time. This is just one example of how a large number of particular fossils in one location can give a false impression of abundance.

## Fish

There would have been a variety of fish living in the rivers and lakes of this area, but by far the most frequently found remains are those of *Lepidotes* (Plate 6). It grew to a length of 60 cm (2 ft), and was covered in hard scales which probably provided a form of protective 'armour-plating'.

As its mouth contained many small, rounded teeth on its upper and lower palates (Plate 4), this suggests that *Lepidotes* fed on aquatic molluscs by crushing their shells. Such observations can provide evidence of a small link in the food web of this environment. In fact *Lepidotes* remains are frequently found together with the fossil shells of bivalves such as *Neomiodon* and *Filosina*.

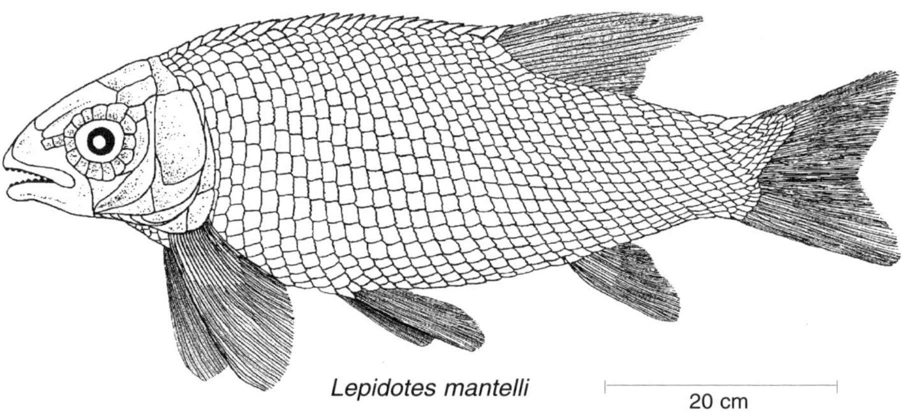


*Lepidotes mantelli*

The teeth of *Lepidotes* were continually growing to replace those which were worn or broken, as they still do in reptiles and sharks today. The modern relatives of this fish include the sturgeon and the bowfin.

The much rarer remains of smaller fish, such as *Neorhombolepis*, *Pachythrissops* and *Leptolepis*, have been found on the coast between Hastings and Cooden Beach. These are occasionally seen in siltstones as a jumble of tiny bones, scales and teeth that have been exposed by erosion.

*Hybodus* was a freshwater shark which grew to a length of 2 m (6 ft). As their skeletons were composed mainly of cartilage, which soon decomposed, only the teeth and dorsal fin spines are generally preserved as fossils. However, in recent years, several ironstone nodules containing hybodont shark skulls have been found in beds of clay on a local beach.

The teeth of *Hybodus* were small but sharp, and each tooth had smaller cusps on either side. Like present-day sharks, *Hybodus* had rows of new teeth inside its jaws with each set slowly moving forwards to replace the worn and broken teeth in front. This continuous replacement accounts for the large number of sharks' teeth found as fossils. From the shape of its streamlined body and pointed teeth, it appears that *Hybodus* was a fast-moving predator. The leading edges of the two dorsal fins were protected by bony spines, which could measure up to 20 cm (8 in) in length. These spines had tiny barbs along their edges, possibly as a deterrent against attack by larger predators or other sharks. The rough skin of *Hybodus* was formed from thousands of very small, pointed scales (denticles). These would have improved water flow over the shark's skin and protected it from parasites and abrasion. Although rare, denticles from the skin of *Hybodus* have been found in the Pett Bone Bed.

On its head the male hybodont shark had two small tooth-like cusps, each of which projected just above and behind the eyes. It appears that these cusps were used in the mating ritual and their size may also have been an indicator of the shark's superiority within the group.

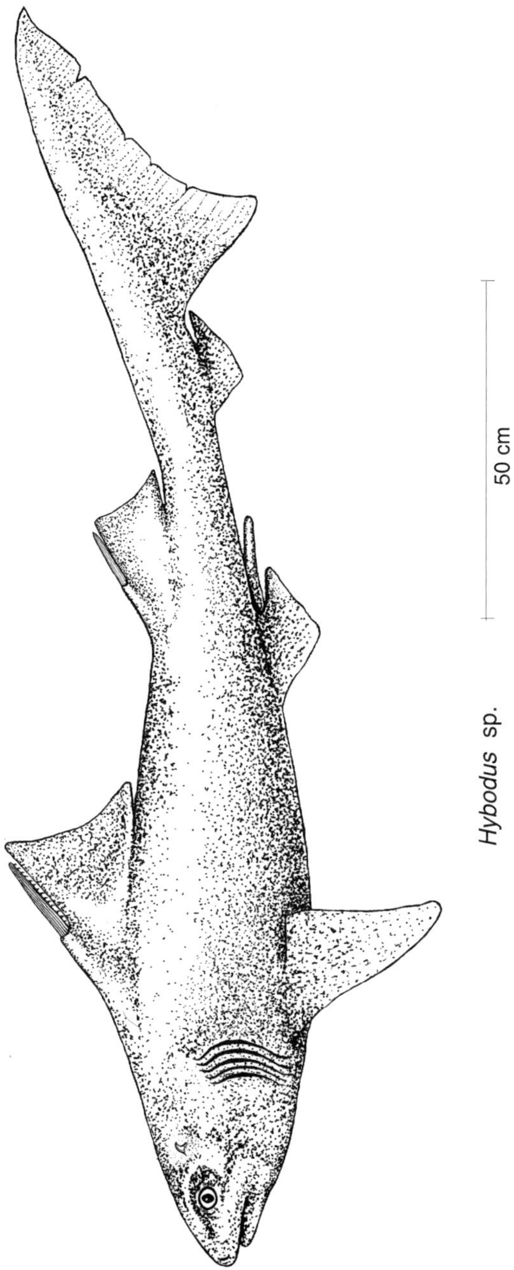


*Hybodus* sp.

## Turtles

The domed upper part of a turtle's shell (carapace) and the flatter lower section (plastron) enclosed and protected its body. The carapace was overlain with horny plates (scutes) which interlocked at the suture lines where growth occured. These animals lived in fresh to brackish water environments and fed on plants. They had parrot-like beaks for cropping vegetation which they ground up with bony ridges on their jaws.

Fossilised bone and carapace fragments from small turtles, including *Plesiochelys* and *Trenosternum*, are occasionally exposed in local rocks. A rare example of an almost complete carapace measured approximately 50 cm (20 in).

## Crocodiles

While dinosaurs were masters of the land, crocodiles dominated the water.

The largest crocodile in this region was *Goniopholis*, a common predator along the shores of the shallow lakes and rivers. In this environment, half-submerged in the water, it hunted for fish and probably other animals such as turtles and small herbivorous dinosaurs. The teeth of *Goniopholis* may be identified by their rounded shapes and parallel striations (Plate 5), which were useful for gripping slippery prey. When part of the upper jaw of this crocodile was discovered on Pett beach, it clearly showed the teeth sockets and a single hole for both nostrils. This restricted nasal passage could be closed by the animal while feeding underwater.

Measurements of its fossil remains suggest that *Goniopholis* may have grown to 4 m (12ft 6in) with a jaw length of 60 cm (2 ft). It was an armoured animal with horny plates under its skin, perhaps for defence against carnivorous dinosaurs - or other crocodiles!

Fossils found in the Bexhill area indicate that there was at least one other species of crocodile living in the lakes and rivers at this time. This was the much smaller *Bernissartia,* which grew to about 60cm (2 ft.) in length and was probably a scavenger. As an opportunist, *Bernissartia* would have ripped through the tough skin or scales of animal remains with its pointed teeth. At the back of its jaws this crocodile had flattened teeth which it may have used to crush shellfish or even bones.

Although both turtles and crocodiles would have left the water to lay their eggs in raised sand banks, one of the great puzzles is the lack of any fossil evidence of reptile eggs or nests in the Wealden area.

**Pterosaurs (flying reptiles)**

Finger bones (phalanges) of the huge pterosaur *Ornithocheirus* have been found near Hastings. It has been estimated from some fossil remains that *Ornithocheirus* may have measured as much as 12 m (40 ft) between wing tips. Despite its great size, it would have weighed under 100 kg (220 lbs) because many of its bones were thin and hollow for maximum strength and minimum weight. Suspended on warm air thermals, this huge flying reptile could glide over 80 km (50 miles) with a single flap of its wings. The vast wing size meant that it could take off from the ground by flapping, perhaps assisted by a light breeze. However, unlike the replaceable feathers of a bird, the stretched skin of a pterosaur's wing would easily have been damaged by twigs and branches.

*Ornithocheirus* hunted for food in coastal areas and lakes where it could swoop down and skim along the water to catch fish with its sharp, pointed teeth. The beak was ideally shaped for this, with keel-shaped upper and lower jaws for stability while cutting through the water.

**Dinosaurs**

The earliest reconstructions of *Iguanodon* depicted it as a heavily-built quadruped, perhaps similar to a rhinoceros in appearance. The life-size models, which were built for the Crystal Palace grounds in 1853, may still be seen today.

Twenty-five years later, after a large number of *Iguanodon* bones were excavated in Belgium, the Mark II version of this dinosaur was re-modelled in a kangaroo-like posture. However, recent studies have shown that *Iguanodon* walked on all four legs for most of the time, but it could also walk or stand on its hindlegs when necessary. Its ability to rise up like this must have been an important advantage in reaching higher vegetation when feeding, and also for spotting approaching predators. While walking on all fours the *Iguanodon's* tail was held horizontally and supported by strong ligaments.

*Iguanodon*

The *Iguanodon* had muscular forearms with front feet that were more like hands. These had hoof-like knuckles for walking, but in place of thumbs were bony spikes which may have been used for defence against predatory dinosaurs. Adult *Iguanodon* skeletons ranging from 6 m (20 ft) to 10 m (32 ft) in length have been found. In most cases these have been identified as separate species, although some authorities claim that they may be male and female of the same species.

By examining the teeth and jaws of extinct animals we can determine what kind of food they ate. In fact, it was the grinding tooth of an *Iguanodon* that originally identified the animal as a plant eater.

The *Iguanodon* was the first dinosaur to evolve a more efficient method of processing food - by chewing it. It had a 'beak' like a turtle for browsing on low-lying vegetation, and two rows of flat teeth on hinged upper jaws. Ferns, cycads and horsetails were difficult to digest but, with grinding teeth, the *Iguanodon* was able to chew tough plants and process food in its mouth. This was a definite advantage over other herbivorous dinosaurs that had to swallow stones (gastroliths) to grind up food in their stomachs. Yet, while these leathery and fibrous plants would soon wear down an animal's teeth, there were always replacements growing in the same sockets. *Iguanodon,* in common with other reptiles, had continuous tooth replacement throughout life. Its beak would also continue to grow, keeping pace with the wear of the tough vegetation.

Dinosaur footprints can provide a great deal of information about the structures and habits of these animals (Plate 7). *Iguanodon* trackways show that its legs were under the body and didn't sprawl out like those of a lizard or crocodile. Sequences of footprints can reveal not only the stride length but also the size of a particular dinosaur. Some bedding planes are covered with overlapping prints, suggesting that a group of dinosaurs may have been feeding in a marshy area. Tracks such as these might even indicate migration routes, where herds left areas stripped of vegetation as they moved on to new pastures. (Grass did not evolve until around 100 million years later!)

When isolated fossil bones are found locally they could have been transported from other areas by rivers in Lower Cretaceous times. However, dinosaur footprints are important because they prove beyond doubt that these animals actually roamed across the muddy flood-plains of East Sussex.

During very stormy weather in November 2000 great quantities of sand and shingle were washed from the beach at Bexhill. This revealed a siltstone bedding plane containing a large number of dinosaur footprints, including short sequences of tracks. Most of these are still filled with the original sediment which now forms casts within the foot impressions.

When surrounded by so many three-toed footprints of varying sizes, it is easy to imagine a family group of dinosaurs feeding on the vegetation by a lake or riverside. Most of the prints have been identified as those of *Iguanodon*, but others are rather unusual. They are smaller, different in shape and have more clearly defined claw marks (back cover). With these observations, and by estimating the animal's size from its 66 cm (26 in) stride, could it be that these tracks were made by a species of theropod dinosaur while stalking a young *Iguanodon*?

Nearby, the siltstone beds contain a number of carbonised tree branches and layers of compressed plant material, perhaps the same vegetation on which the dinosaurs were feeding!

*Iguanodon* footprints are fairly common in the south-east but, as those of predatory dinosaurs are rare, there is insufficient evidence to suggest that they hunted in packs. It is possible that the herbivore/carnivore ratio in Lower Cretaceous times may have been similar to that of present-day species; for example, on the African plains there are hundreds of wildebeest to every lion.

Geologists used to believe that a great 'Wealden Lake' extended over most of southern England and across to central France; but such a huge lake would have left insufficient land to support the herds of *Iguanodon* that apparently roamed this region. With vegetation restricted to the margins of a single lake, the plant cover would soon have been eaten by migrating dinosaurs. Also, the sedimentary structures in the sandstones of this area are typical of those laid down by rivers flowing across flood-plains and deltas into smaller lakes.

In the past nearly all of the large bones and footprints found in the Wealden district have been identified as those of *Iguanodon*, but there is now exciting evidence of other dinosaurs.

Recent dinosaur discoveries in the Bexhill area have included spines, bones and scutes which have come from an armoured dinosaur, perhaps similar to *Hylaeosaurus*. This well-protected, plant-eating animal was covered in bony plates with strong spines on its back and tail. It might well be described as the 'tank' of the dinosaur world

There is also a vertebra from a fast-moving predator, possibly a dromaeosaur, which was about the size of an average dog. This theropod ran on its back legs and used sharp, sickle-like claws to kill and cut open its prey. With large eyes and binocular vision, the dromaeosaur may have hunted small mammals at night.

Unfortunately, articulated skeletons are very rare because most animal remains would have been torn apart by scavengers, then perhaps washed into rivers or lakes by rainstorms. The winnowing process of the flowing water would eventually deposit the bones, scales and teeth some distance apart.

One of these rare discoveries was made in 1983 by Mr. William Walker, an amateur fossil collector. He was searching for specimens in a Surrey clay pit when he found a huge claw in a nodule of siltstone. The claw, measuring 24 cm (10 in) in length, belonged to a

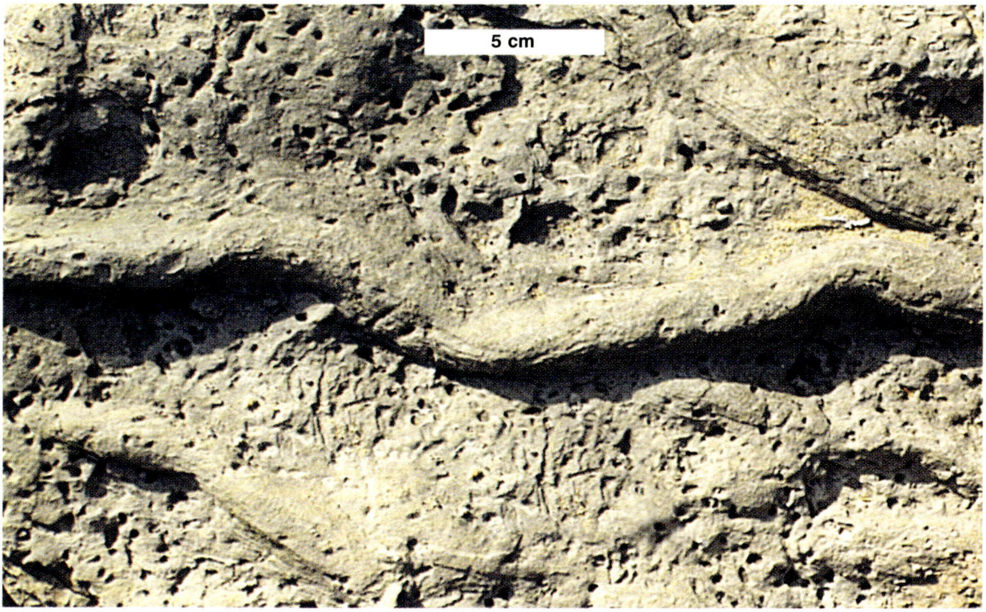


Plate 10. Gutter casts

Plate 11. Infilled stream channel

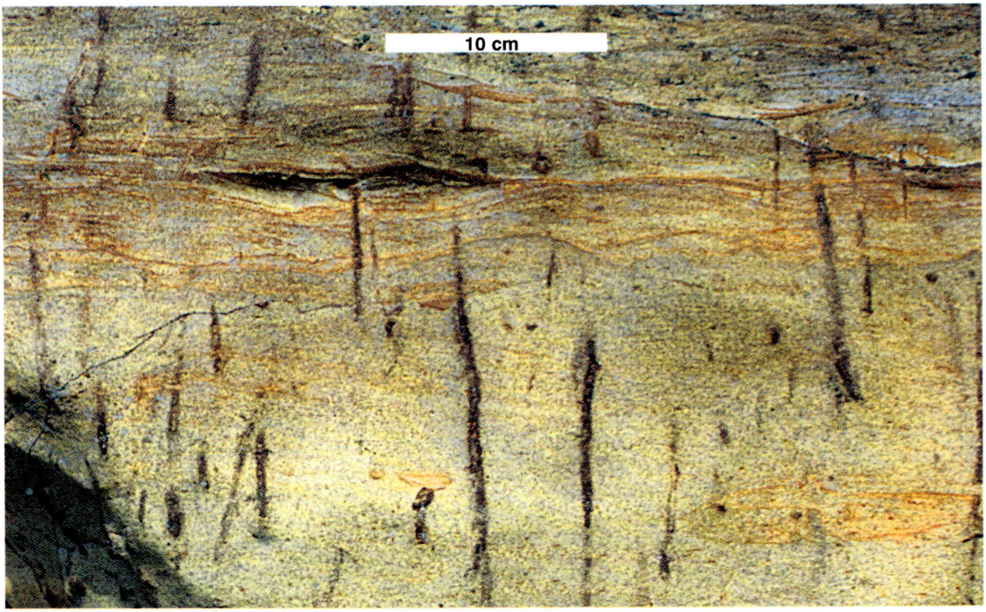


Plate 12. Horsetail roots: *Equisetites*

Plate 13. Bennettitalean: *Pterophyllum*

large and unknown carnivorous dinosaur which was later named *Baryonyx walkeri* in honour of its discoverer.

When reconstructed, it was calculated that the dinosaur weighed about two tonnes and was about 10 m (32 ft) long. The animal walked on its back legs and had a crocodile-shaped skull with 64 teeth in its lower jaw, twice the number found in other theropods! Recent research indicates that this dinosaur was probably a hunter and a scavenger, with large claws for ripping open a carcass. Semi-digested scales of *Lepidotes* found in its stomach cavity provided evidence that *Baryonyx* was also a fish-eater.

Before the discovery of *Baryonyx,* no almost complete large theropod had been found in Lower Cretaceous rocks anywhere in the world. We know that this dinosaur lived locally, as a single tooth from Baryonyx has been found in the Wadhurst Clay of the Bexhill area. It is similar to a crocodile's tooth, but has finely serrated cutting edges.

# CONCLUSION

There is still a great deal to be discovered and learned about life in the Lower Cretaceous period. All of the fossil species discussed in this book may represent only a fraction of those which probably existed at that time. However, the local rock structures and fossils still have a very important contribution to make in any reconstruction of ancient habitats and communities. Also, an insight into how extinct plants and animals once lived may be gained by comparing them with similar present-day species.

When all of this is considered, a field trip can become a journey back to East Sussex as it was over 100 million years ago. We can imagine a landscape of meandering rivers and marshy flood-plains with sub-tropical vegetation and an abundance of animal life.

There is much enjoyment to be gained in geology simply by observing, recording and trying to understand or explain what you have seen. It is hoped that by promoting the study and interpretation of local geology and fossils, a greater appreciation of present-day environments and their importance may be achieved.

A visitor to this area once summed up the fascination of a geological field trip with the comment: 'From now on a walk along the beach will never be the same!'

# WHAT ARE FOSSILS ?

Fossils are the preserved evidence of prehistoric life. They are the clues that are used by palaeontologists to study the evolution of species and to reconstruct ancient communities. There are certain conditions which are essential for the preservation of animal remains:

## BODY FOSSILS (Moulds & Casts)

The bodies of dead animals soon decay or are eaten by scavengers. If the remains are deposited in water they may gradually become buried under layers of sand, mud or silt. Over millions of years, as these sediments are consolidated, the bones, teeth or shells are sometimes dissolved by chemicals in groundwater, leaving hollow moulds or impressions. Eventually these are filled by sediments or minerals that crystallise from groundwater solution to form casts (Plates 1 and 2).

## REPLACEMENT (Diagenesis)

In many fossils the internal structures may also be preserved. The original minerals are gradually recrystallised or replaced by other minerals from solution. These include calcite ($CaCO_3$), quartz ($SiO_2$) and iron pyrite ($FeS_2$).

## CARBONISATION (Plants)

Plant material is compressed to a thin layer of carbon. Carbonised fossils are commonly found in siltstones and the roof shales of coal mines. Examples: ferns (Plate 13) and wood.

## TRACE FOSSILS (Indirect evidence)

These may be the only surviving clues to an animal's past existence. Examples: Bioturbation (burrows), coprolites (faecal pellets), teeth marks and footprints (Plate 7).

**PSEUDO FOSSILS (False fossils)**

Features in rocks which look like fossils. Examples: conchoidal (bivalve-shaped) fractures in flint, dendritic (plant-like) patterns formed where minerals crystallise in rock fissures, and natural concretions which look like fossilised eggs.

In 1921, stone fragments of a strange structure were discovered in the Wadhurst Clay during road works in Hastings. When reconstructed, they were identified as parts of the internal cast of a gastropod, measuring about 2 m (7 ft) in length, and it was given the name *Dinocochlea ingens* (terrible/huge unknown snail). Evidence of an organic origin was suggested by its regular spiral, although there was no trace of any external or internal shell material.

*Dinocochlea ingens*

25 cm

Since then, the true nature of the structure has been discussed and argued at length. It was even claimed to be the burrow of an extinct beaver (in rocks dated at 135 million years old?) Despite all the controversy, it is probably a natural sandstone concretion which just happened to form in the shape of a large gastropod. There are still those who believe that *Dinocochlea* is a rare fossil species of mollusc but, to date, no further examples are known to have been found.

This illustrates how difficult it can sometimes be to distinguish a genuine fossil from a pseudo, or even a faked specimen.

# GLOSSARY

**Bioturbation**

Where sediment has been disturbed or burrowed by any kind of animal, or by plant growth.

**Braided streams**

Streams that divide into numerous branching channels, which often re-join to form small islands with sand bars between them.

**Cement**

A geological term for any mineral which crystallises between particles of sediment to form a rock. Calcite is a very common cement.

**Concretion**

Hardened lumps or nodules found in shales, sandstones and limestones. They are usually formed by the precipitation of minerals around a central nucleus, such as a shell fragment, within the sediment. Concretions may be rounded or irregular in shape.

**Conglomerate**

A sedimentary rock with a fine-grained matrix around poorly sorted material ranging from granules to pebbles which were rounded by transport in water.

**Fault**

A fracture which occurs in rocks where the strata have been displaced as a result of stretching (normal fault) or compression (reverse fault).

**Flash flood**

Where a heavy storm on the high ground causes a great volume of water to pour down a river channel and across its flood-plain.

**Flood-plain**

The area on either side of a river where mud and silt (alluvium) are deposited when it overflows during floods.

**Fusain**

Wood that has been turned into charcoal, usually by forest fires.

**Gastroliths**

Stones swallowed by dinosaurs to grind up fibrous plant material in their gizzards. Also known as 'stomach stones'.

**Igneous rock**

A rock that has solidified from molten material (magma) which was generated deep within the earth. Examples: basalt, granite.

**Lens**

A lens-shaped rock formation or sedimentary deposit which lies horizontally between bedding planes, and narrows to a point at both ends.

**Precipitate**

In chemistry, the deposition of solid particles (crystals) out of a super-saturated solution.

**Prograding delta**

Sediments deposited by rivers on a shore or lake side form deltas which gradually advance (prograde) outwards into the sea or lake.

**Rhizomes**

Underground stems which produce roots and leafy shoots. Example: horsetails.

**Scavenger**

A meat-eater that feeds on dead creatures and steals food from other animals.

**Scree**

Broken rock fragments at the base of a cliff or steep incline. The term 'talus' is used specifically for rock debris below a cliff.

**Scutes**

A horn-like material forming exoskeletal plates which cover and protect the shells of turtles. In crocodiles, the scutes are hard plates embedded in the skin.

**Sp.**

An abbreviation of 'species'.

**Theropod**

A bipedal, carnivorous dinosaur.

**Truncated laminations**

Sedimentary bedding planes that have been cut off abruptly by erosion or faulting.

# FURTHER READING and REFERENCE BOOKS

**The Wealden District.** (British Regional Geology).
F. H. Edmunds. H.M.S.O., 1954

**British Mesozoic Fossils.**
The Natural History Museum, 1975

**The Weald: A Geological Field Guide.**
W. Gibbons. Unwin Paperbacks, 1981

**Geology of the Country around Hastings & Dungeness.** (B.G.S.).
R. D. Lake and E. R. Shephard-Thorn. H.M.S.O., 1987

**The Practical Geologist.**
D. Dixon. Aurum Press, 1992

**Early Cretaceous Environments of the Weald.** (Guide No.55).
A. Ruffell, A. Ross and K. Taylor.
The Geologists' Association, 1996

**Rocks and Fossils.**
A. Busbey, R. Coenraads, D. Roots and P. Willis.
Harper-Collins, 1996

**Food of the Dinosaurs.**
B. A. Thomas and C. J. Cleal.
The National Museum of Wales, 1998

**Walking With Dinosaurs.**
The Natural History Museum, 1999

# MUSEUMS AND GEOLOGICAL GROUPS

Examples of local rocks and fossils can be seen at:

**Hastings Museum**, Cambridge Road, Hastings 01424 781155
**Bexhill Museum**, Egerton Road, Bexhill 01424 787950
**Maidstone Museum**, St. Faith's Street, Maidstone 01622 754497
**The Booth Museum**, Dyke Road, Brighton 01273 292780
**The Natural History Museum**, Cromwell Road, London.
0207 942 5000

Further information may be obtained from:

**Hastings and District Geological Society**
Contact: Ken Brooks 01424 426459

**West Sussex Geological Society**
Contact: Pat Waterston 01903 265715

**Brighton and Hove Geological Society**
Contact: John Cooper 01273 292780

**Horsham Geological Field Club**
Contact: Gill Woodhatch 01403 250371

**Sussex Mineral and Lapidary Society**
Contact: John Pearce 01444 233958

**The Geologists' Association**
Contact: Sarah Stafford 0207 434 9298

A trip to the **Hastings Country Park** and its Visitor Centre is certainly worthwhile.

# INDEX

Page numbers in **bold type** refer to illustrations and diagrams